China Video Service Experience

Authored by: China Video Service User's Experience Standards Working Group

人 民 邮 电 出 版 社
北 京

图书在版编目（CIP）数据

中国视频服务体验 = China Video Service Experience : 英文 / 中国视频服务用户体验标准工作组编著. -- 北京 : 人民邮电出版社, 2017.4
ISBN 978-7-115-45320-4

Ⅰ. ①中… Ⅱ. ①中… Ⅲ. ①视频系统－商业服务－中国－英文 Ⅳ. ①TN94

中国版本图书馆CIP数据核字(2017)第061257号

◆ 编　　著　中国视频服务用户体验标准工作组
责任编辑　李　静
执行编辑　贾朔荣
责任印制　彭志环

◆ 人民邮电出版社出版发行　　北京市丰台区成寿寺路 11 号
邮编　100164　　电子邮件　315@ptpress.com.cn
网址　http://www.ptpress.com.cn
北京瑞禾彩色印刷有限公司印刷

◆ 开本：787×1092　1/16
印张：5　　　　2017 年 4 月第 1 版
字数：55 千字　　　　2017 年 4 月北京第 1 次印刷

定价：68.00 元

读者服务热线：(010) 81055488　印装质量热线：(010) 81055316
反盗版热线：(010) 81055315
广告经营许可证：京东工商广字第 8052 号

Editorial Board Member

China Video Service User's Experience Working Group

Member Units

China Academy of Information and Communications Technology

Academy Of Broadcasting Planning,SAPPRFT

China Telecommunications Corporation Ltd.

China Mobile Communications Corporation

China United Network Communications Group Co., Ltd.

Huawei Technologies Co., Ltd.

Beijing iQiYi Science & Technology Co., Ltd.

Tencent Video

Youku Tudou Inc.

Shanghai Jiao Tong University

Wangsu Science & Technology Co,.Ltd

Beijing Bohui Science & Technology Co., Ltd.

Special Thanks

BeiJing BONREE Data Technology Corp., Ltd

TingYun

Preface

With the rapid construction of broadband and fixed network infrastructure and the development of mobile Internet services, various video services have emerged in recent years. Video services, such as live sport event broadcasting and TV series tracking, are becoming essentials to people's information consumption in all aspects, from entertainment to study and communication.

As video services are flourishing, consumers have higher requirements regarding definition, multi-screen display, interactivity, sociality, and instantaneity. These requirements pose great challenges to video operators and telecom carriers. Under these circumstances, a widely-accepted video experience assessment standard must be set up so that video operators can develop and provide content that meets consumer requirements while carriers can expand the video service market.

Under the joint efforts of the parties involved in the group working to develop this standard, the first book about the video experience assessment standard is released in China. The standard not only provides comprehensive and systematic suggestions on video service standardization, but also dissipates consumers' concerns. In addition, the standard provides rich practice references for network pipe optimization from the video service perspective and standardizes the commercial operation of video services.

The *China Video Service Experience* is jointly written by Yang Kun, Li Jing, Chen Jincong, Hu Xianghui, Tang Xin, and Zhang Yutao. Here, we would like to thank the authors for their efforts, our colleagues for their great contributions to this book, and the Post & Telecom Press for their support.

We hope that this book can be of benefit to readers and make a modest contribution to the development of the video industry.

Editorial Team

Content

Video services are under rapid development throughout the world. They have greatly pushed broadband construction and network acceleration and have become a key measure for optimizing ICT resource allocation and promoting ICT industry transformation. Driven by video applications, video communication and video surveillance services have also grown rapidly. They form a huge video business system, which is becoming the foundation of smart city and advanced manufacturing.

User experience greatly affects consumers' selection of video services. Video services require real-time, continuous, stable, and E2E controllable streaming transmission, which is a huge challenge for the Internet that features great traffic changes and resource sharing. Although telecom carriers continually expand bandwidth, yet users' video experience has not been effectively improved. The key reason is the lack of a unified video experience assessment standard. As a result, network layer's influence on continuous and stable video streaming transmission can't be precisely evaluated.

Service providers who fail to provide satisfied services to users can't keep them. To ensure sustainable development of the video industry, leading companies and organizations, including China Academy of Information and Communications Technology (CAICT), State Administration of Press, Publication, Radio, Film and Television (SARFT), Academy of Broadcasting Planning, China Telecom, China Mobile, China Unicom, Huawei, Youku, Tencent Video, iQIYI, Shanghai Jiao tong University, BROADV, and Wangsu Science & Technology, jointly set up a working group and published the first *Video Experience Assessement Standard (1.0)*. The standard aims at providing the best video experience to users and pushing the healthy video industry development.

The working group also invited some work units to evaluate users' experience of IPTV and Internet video services in some areas in China to verify the effectiveness and practicability of the Standard. The evaluation shows as follows:

- The experience of different video services in China varies greatly and there still has plenty room for improvement.
- Viewing experience of IPTV services based on private IP networks is great. However, video quality and interactive experience still have to improve. In general, the overall experience in this scenario is favorable.
- Internet video services use shared network resources with other services. Users' experience in this scenario is adversely affected and need to be improved.

- For internet video services, users' experience on mobile phones is better than that on PCs'.
- Users' experience of video services provided by different carriers and content providers (CPs) varies.
- 4K video services need to do more to improve users' experience than HD and SD video services do.

This work lays the foundation of constructing a comprehensive video service experience evaluation system in China in the future. When the effectiveness and practicability of the standard are verified, the standard will be gradually applied to all video play systems of units involved in the working group, then further cover the tremedous video business system including video communication, video conference, video surveillance, video sharing, and digital signage services, finally align with the international standards, filling in gaps of the international video industry.

1. Services are Developing Rapidly with Network Construction and Technology Improvement

The Government attaches great importance to broadband infrastructure development. The 13th Five-Year Plan for National Economic and Social Development proposes to quickly construct the new generation high-speed, mobile, secure, and ubiquitous ICT infrastructure. Pushed by the Broadband China strategy, the ICT industry accelerates the construction of broadband networks and mobile broadband (MBB) networks. According to the data published by the Ministry of Industry and Information Technology (MIIT), the total number of broadband subscribers of China Telecom, China Mobile, and China Unicom reached 292 million and that of MBB subscribers (3G and 4G mobile users) reached 885 million by the end of September, 2016. Both the scale of users and network speed increase greatly. Video services play a vital role in driving up the network construction and speed.

1.1 Video Service Becomes the Main Contributor of Network Traffic

Traditional TV services such as cable TV, satellite TV, and terrestrial TV services are growing slowly. On the contrary, pushed by the Government-led practical exploration of networks convergence, IPTV services, online Videos, and mobile Videos are flourishing. By the end of 2015, more than 20% of pay-TV users were served through broadband networks worldwide. Mobile video services also develope rapidly. In China, the main video services such as IPTV, cable TV, and Internet TV services are all growing.

- The number of IPTV users of China Telecom, China Mobile, and China Unicom increased sharply in 2016. Mobile video users in China increase quickly in recent years and are expected to reach 527 million.
- China already has 254 million cable TV subscribers. 200 million users have completed TV digitization transformation; over 50% users are served by two-way HFC networks, and over 20 million users are wired-broadband users.
- As reported by China Internet Network Information Center (CNNIC), online Internet video users reached 514 million by June, 2016. All main

video CPs had deeply participated in content creation and production. High-quality content further helps users to form the habit of paying. By 2015, there were 165 million Internet video terminals (including smart TVs and STBs) used. According to the research carried out by Nielsen-CCData, the number of annual active Internet video users is between 27 million to 30 million in China.

With the development of video services, video traffic has accounted for 75% of total IP traffic. Till 2020, the compound annual growth rate (CAGR) of global Internet video traffic is expected to be 25%, as shown in Figure 1-1. In 2020, video traffic on mobile networks will account for 18% of global internet video traffic, with the highest CAGR, which is expected to be 65%. The proportion of video traffic on Wi-Fi in global internet video traffic will reach 50% with a CAGR of 24%.

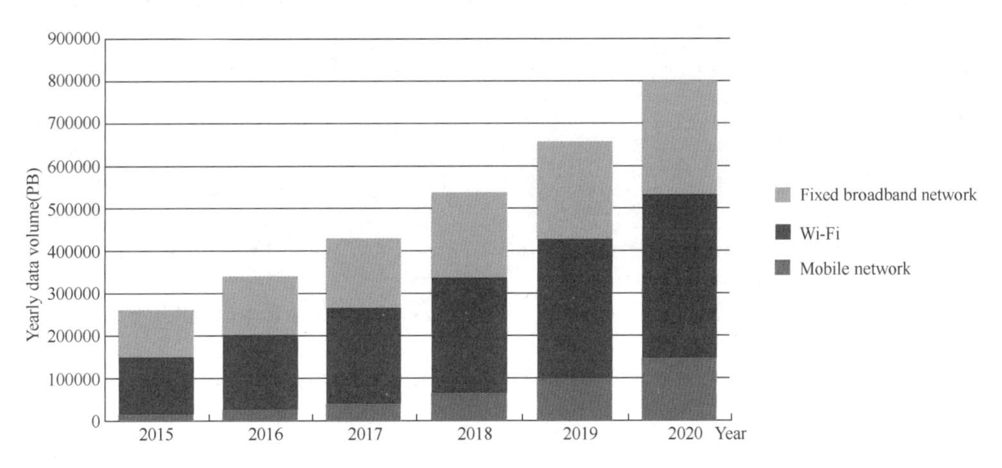

Figure 1-1 Global Internet video traffic forecast

1.2 Video Service Is Becoming a Major Driver for Network Acceleration

Broadband networks and video services support and promote each other and associate closely. Telecom carriers such as China Telecom and China Unicom have regarded video service as the foundation telecommunication service. The development of video service has become a core factor affecting the development of the entire broadband industry.

- Video service is the only choice to fill the bandwidth of accelerated fix and mobile broadband networks, meanwhile bundles with broadband service. They can boost users' demand of bandwidth consumption. According to statistics from carriers, the optical broadband penetration rate of IPTV services has exceeded 50% and the proportion of active users has exceeded 70%.
- Video sevices directly raise carriers' revenues. For example, the growth rate of AT&T video services in 2015 was 196.9%, which contributed 93.6% of the total revenue growth rate. In China, the broadband VAS revenue brought by video also grows quickly as reported by IPTV carriers.
- The sales of 4K video terminals keep increasing. Most video CPs are considering the deployment of 4K video services. China Telecom, China Mobile, and China Unicom deployed more than 45 million 4K IPTV

STBs in 2016. Widespread use of 4K video services requires stable and high-speed broadband networks to guarantee the high-quality streaming transmission. At least 30 Mbit/s bandwidth is required for 4K video playback and at least 50 Mbit/s stable bandwidth is required for smooth 4K video playback. Therefore, 4K video service is a major force for future broadband upgrade.

- Video CPs all over the World research the next generation video services such as 8K, VR and AR. In some areas of China, plans have been already proposed to implement Gigaband to serve millions of users in 2018, the average access bandwidth will increase from 50 Mbit/s to 280 Mbit/s, and the perceptible download rate will increase from 13 Mbit/s to 100 Mbit/s. The next generation video services will still play an important role in driving network acceleration.

1.3 Video Service Is a Major Factor to Accelerate Industry Transformation

Video service is closely associated with network upgrade and becomes the mainstream of the ICT industry. It brings user growth and pushes industry transformation.

- Video play is a key service that is closely bundled with network resources of carriers. It protects the network value of telecom and cable TV carriers against the challenge of competing with Internet video CPs, it also enables carriers to keep their dominance and flow control rights, and pushes Internet video CPs to cooperate deeply with carriers.

- Live video is users' rigid demand. Service platforms can use content to improve user traffic. Multi-screen interactive videos can always attract users. With the flow control rights of video play, carriers or CPs can acquire traffic resources of other services on the same service platforms.
- CPs must have the end-to-end operational capability, including content production, network distribution, service aggregation, content storage, data operation, and terminal video playback, to provide users with a satisfactory experience of video service. Video service development drives all CPs to construct a service ecosystem with integrated operation capabilities from content to terminals, also from services to networks.

Global telecom carriers, cable TV carriers, and Internet video CPs are seizing this opportunity by integrating industrial chain resources, constructing integrated operation capabilities, and transforming the development model. The close integration of industry chain can bring unprecedented cooperation of manufacturers in the aspects of traffic monetization, channel management, and content marketing. Bundling sale becomes the developing trend of industry chain. National carriers and CPs are cooperating to upgrade the IPTV and the Internet video platforms to achieve the goal.

1.4 Video Service Is Now Extending Its Application to a Wider Range

Based on the high-speed fixed mobile convergence broadband networks and large user scale, carriers can use capabilities such as smart networking, rich intelligent peripherals, open data and cloud platforms to provide HD and multi-screen video applications for users. A complex video business system associated with broadband networks is on its way.

- Video communication is widely spreading with the development of SNSs, multi-screen video services and 4G networks. An increasing number of users get used to communicating by videos on mobile phones, tablets, computers, TVs or game consoles. Unlike the traditional voice service, video communication will be merged as a basic function into other services, such as SNSs, e-commerce, online education, and online games.
- More and more enterprises integrate video communication and video conference into their IT platforms to provide flexible, HD, and smart contact methods.
- As a key enabler of Internet of Things (IoT) development, traditional video surveillance service gains a new opportunity and is widely used in various fields such as production management, city security, traffic commanding, and agricultural production. In addition to video information

collection, smart HD video surveillance services can intelligently process video information after connected to the backend IT platform, big data analytics platform, and video playback platform. The camera resolution for video surveillance becomes more smart and more detailed information can be collected. Combined with the technologies such as big data analysis and facial recognition, video surveillance can provide further intelligent services.

- With the support of backend networks and cloud platforms, digital signage associates large outdoor screen, IoT, and mobile Internet technologies to implement information publication and user interaction, which fundamentally changes the traditional advertising and marketing modes.

The comprehensive video business system is penetrating all aspects of social life. Currently, the smart home ecosystem based on broadband networks and multi-screen video draws the most attention. The system will be extended to smart communities and finally smart cities. In addition, deep combination of the next generation IT and manufacturing brings new production modes, industry forms, and business models, where video surveillance, video communication, video navigation, and video conference play important roles. Different levels of capabilities in the big video business system are gradually formed based on broadband and mobile networks and will be widely used in various fields such as administration office, enforcement, agricultural production, manufacturing, energy and power, financial transaction, travel, transportation and logistics, education, and healthcare.

1.5 Developing Trend of Video Services

Video service is developing toward a comprehensive trend involving movies, games, shopping, and communication. Multi-screen, HD, interactive, and social content is now becoming the core of video consumption. The large video service platform with the large screen be regarded as the basic entry, content and service as the user entry, and O2O e-commerce as the offline entry will be a main convergence point of the future Internet traffic. Its influence will exceed current Internet portals'. Because the lack of flexibility in capability upgrading, current video service platforms need to be improved to a huge video business system to meet the following business development requirements.

- Business openness and resource scheduling scalability. Existing video system architecture needs to be optimized and decoupled to support the rapid-developed new business generation. Modules could be assembled flexibly and dynamically.
- Flexible deployment network. Using multi-screen fusion service, fixed mobile convergence network can provide services to all mobile and fixed network users. This can be implemented by deploying fusion CDN in the beinging. Finally, smart virtual communication pipelines are required to

meet business needs.

- Combination of cloud computing, CDN, and smart terminals. Cloud computing provides on-demand network resource pool for users. CDN acceleration can effectively optimize content transmission. Combination of cloud computing and CDN can make flexible service traffic distribution come true. Together with smart terminals, services can be quickly deployed and flexibly configured.
- E2E big data support. Big data helps to improve user experience and ensure QoS. Users can be provided with precise and personalized services such as intelligent search, personalized recommendation, personalized advertising, as well as personalized stores based on user profiles.
- New marketing pattern. With the support of platform's flexible capabilities and big data, video e-commerce and community life services and be expanded, new marketing modes such as interactive and targeted advertising can be developed.

2. Growing Demand for Unified Video Experience Assessment Standards

2.1 Video Experience Determining Users' Selection of CPs

The image dynamic range that human eyes can observe is much wider than that constructed by the current technology. Users have higher requirements for content variety and image quality. Based on the network survey, 71% of Chinese users emphasize that HD image quality is a key factor affecting their choices, 65% of users emphasize that CPs providing 4K/UHD image quality draw more attention from them, and 45% of users are willing to pay for 4K/UHD videos. Previously, users selected carriers mainly based on the web page response time or software downloading speed. Currently, more users select carriers based on the video viewing experience. Therefore, constructing an experience-oriented network is a key issue for carriers to improve their competitiveness.

As surveyed by Ovum, 28% of respondents said they changed their broadband carriers within the past 12 months, among whom 64% of respondents changed the carriers because of faster network speed or cheaper tariff (or both). Besides, more users judge the network speed based on their video playback experience. Users' experience greatly affects user's selection of video services. Video experience requires real-time, continuous, stable, and E2E controllable streaming transmission. Figure 2-1 shows the proportion of users who wanted to change their broadband carriers within the past 12 months in all respondents and are undergoing good or poor media service experience. Figure 2-1 also shows that those with poor media service experience are double than those with good experience.

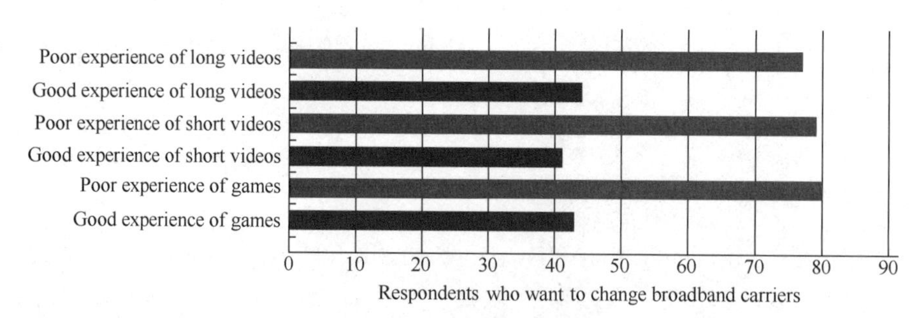

Source: Ovum

Figure 2-1 Doubled carrier change probability for users with poor video experience

The survey also shows that the key factor for users to select video CPs is their video experience. As surveyed on networks by XXX (telecom carrier), unsmooth video playback is the top one factor affecting IPTV user activity, as shown in Figure 2-2 and Figure 2-3.

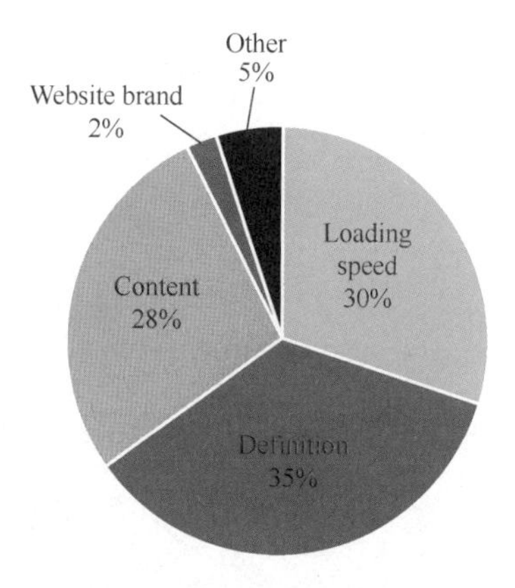

Figure 2-2 Factors affecting users' selection of video websites

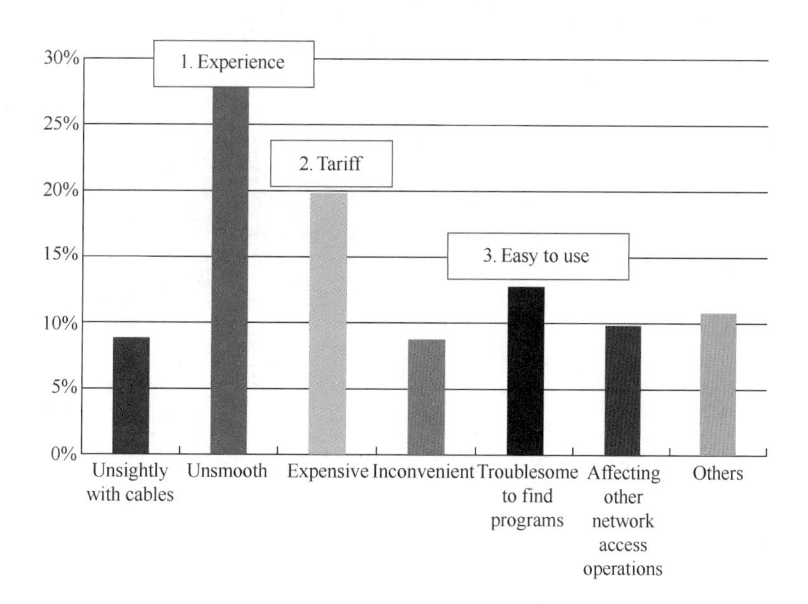

Figure 2-3 Main factors affecting IPTV users' activity——surveyed by XXX (telecom carrier)

As a conclusion, video experience determines users' satisfaction and activity. HD and smooth video playback are key factors affecting users' selection of CPs.

2.2 Multiple Measures for Improving Users' Experiences

The entire video industry is continuously striving to provide optimized video experience. As technology progresses, the mainstream resolution for video services on the market has evolved from SD to HD and 4K, and will migrate to higher levels (8K, AR, or VR) in the future. Video CPs are taking various measures to improve users experiences. Gradual improvement of video experiences is demonstrated in Figure 2-4.

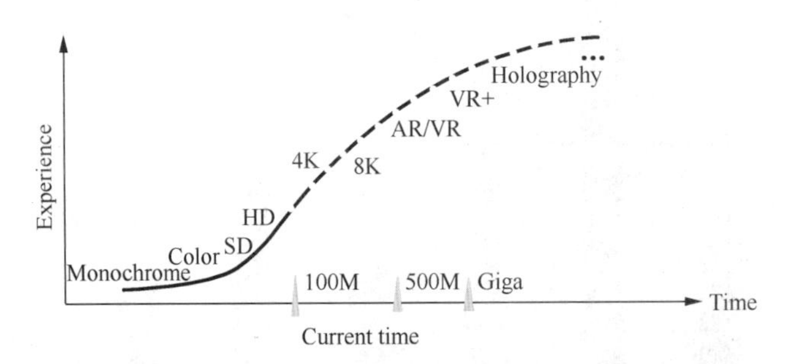

Figure 2-4 Gradual improvement of video experience

2015 was the first year of the 4K video era. Global chargeable TV carriers provided 4K TV services one by one. Carriers in China have migrated from SD to HD video services and began to provide 4K video services in multiple provinces. China Telecom has provided 4K video services on the IPTV platform in multiple provinces and is further promoting ultra 4K applications. On the basis of 3840 × 2160 resolution, the goal is to raise the video frame rate from 30 frames per second (P30) to 60 frames per second (P60), to increase the color depth from 8 bits to 10 bits, and to improve the image lightness from standard dynamic range (SDR) to high dynamic range (HDR). China Unicom, defining the all-optical network (AON) era as the 4K video era, it has established a China Unicom 4K laboratory, released standards and specifications for the full video service system involving services, products, networks, platforms, and terminals, the laboratory also tested and implemented wide coverage of 4K video services. China Mobile has already released 4K STBs and is now constructing 4G networks that can provide excellent video experiences in multiple provinces. SARFT is also deploying 4K TV services on cable networks in multiple provinces.

Internet video CPs in China are also making efforts to improve users' video experiences. Since 2012, iQIYI has increased its average bit rate by about 30% and invested a lot trying to reduce rebuffering frequency and duration during video playback. The proportion of rebuffering duration decreased from 6% or 7% in 2012 to 1% in 2016 and aimed to approach 0% finally. Youku has been focusing on improvement of the video bit rate and resolution by upgrading the bandwidth and using the U+ patent technology. In addition, Youku improves the video definition by over 50% using measures such as flexible scheduling, smart adaptation and smooth switchover. Tencent Video has constructed a video quality monitoring system to report areas or users with poor video quality and conduct targeted upgrade to improve the overall users' experience.

2.3 Lack of Unified and Effective Video Experience Assessment Standards

The traditional method for evaluating the video service quality is to simulate the actual environment in a lab or collect broadband network indicators to estimate the service quality. However, unlike other Internet services, video services, especially entertainment video services, which depend more on the valid bandwidth that users actually use, have a high requirement for timeliness, and can be easily affected by network features such as packet loss, delay, or jitter. Therefore, the actual users' experience is hard to measure by simulation or estimation methods. Currently, the information obtained from the deployed labs or broadband network monitoring is incomplete and can't effectively reflect the video service quality.

- A lab environment can only be used to evaluate the video service quality in static scenarios. However, the actual network environment is dynamic and affects services a lot. The result obtained in a lab environment can't be used to estimate the live network situation or locate live network errors.
- The traditional broadband network indicator–based troubleshooting method can't measure users' experience or locate errors in time. For example,

the broadband downloading rate shows the average data transmission level in a period. According to the published data, the average downloading speed (with busy and idle hour weights) of fixed broadband users was 11.03 Mbit/s in the third quarter of 2016, which was sufficient for 8 Mbit/s HD video transmission. However, there are still frame freezing issues, artifacts, or even black screen based on users' feedback. What causes the deviation is that video service experience is determined by the transient comprehensive network and service performance at each specific time instant instead of the average performance in a period. The transient performance is affected by various factors scattered in the entire E2E process and is dynamic. No carriers can obtain all transient performance indicators immediately to estimate users' experience in time. As a result, how to evaluate the real-time video service experience is a big challenge for all carriers throughout the world.

- Deep cooperation between different manufacturers in the industry becomes a trend, but these manufacturers use different methods to detect and analyze factors affecting users' experience. Standards are required to unify the factors affecting users' experience. Early in 2009, ITU-T started to work on video quality assessment standards. In 2012, ITU-T published the first video quality assessment standard, considering mainly network factors but not comprehensive factors in the entire E2E process. The standard has limited effect for evaluating video service experience and instructing video experience design and optimization.

To resolve the difficulty, the industry must carry out unified video experience assessment standards, and establish a monitoring, analysis, and indicator system covering the full process including both users and contents. Using the unified standards, manufacturers can precisely evaluate video experience and quickly locate faults in various scenarios and on different networks. The standards must be executable, extensible, and synchronous with network and service development.

Unified video experience assessment standards and measures have the following important values.

- Providing a scientific reference for administrations to conduct industry supervision and formulate industry policies as well as for local governments to make local smart city plans.
- Providing a progressive QoS evaluation measure for CPs and carriers to analyze the effects of networks, platforms, and terminals on QoS, to locate core reasons, and to improve QoS.
- Helping CPs to deploy "broadband + big video" services differentiated for specified user groups and develop differentiated competitive advantages in the video era.
- Offering an objective basis for manufacturers to optimize their products and platforms, accelerate service innovation, and enhance service effectiveness.
- Providing objective and effective evaluation data for users to select video CPs, protecting their right to know and select.

3. Joint Research of Video Experience Assessment Standards and Onsite Sampling Evaluation

At present, it is difficult to measure video service quality, locate the main influence factors during video consumption, thus users' rights can't be guaranteed. To solve these problems, leading companies and organizations in the video industry, including CAICT, SARFT Academy of Broadcasting Planning, China Telecom, China Mobile, China Unicom, Huawei, Youku, Tencent, iQIYI, Shanghai Jiaotong University, BROADV, and Wangsu Science & Technology, jointly set up a working group for China video experience assessment standards. The working group aims at providing the best video experience for users and pushing healthy video industry evolution, it published the first *Video Experience Assessment Standard (1.0)* on September 27, 2016. The publication of the standard fills a gap of the video industry throughout the world. The standard specifies the estimation models and key parameters for video evaluation, and can be expanded to video communication, video conference, and video surveillance services. It provides solid guarantees to the evolution of the big video industry.

3.1 Main Factors of Video Experience

After interviewing video CPs and users as well as analyzing large numbers of samples, the working group finds that compared with traditional factors such as content diversity and content novelty (as shown in Figure 3-1), the following new factors have more evident influence on user experience of video playback at the present stage.

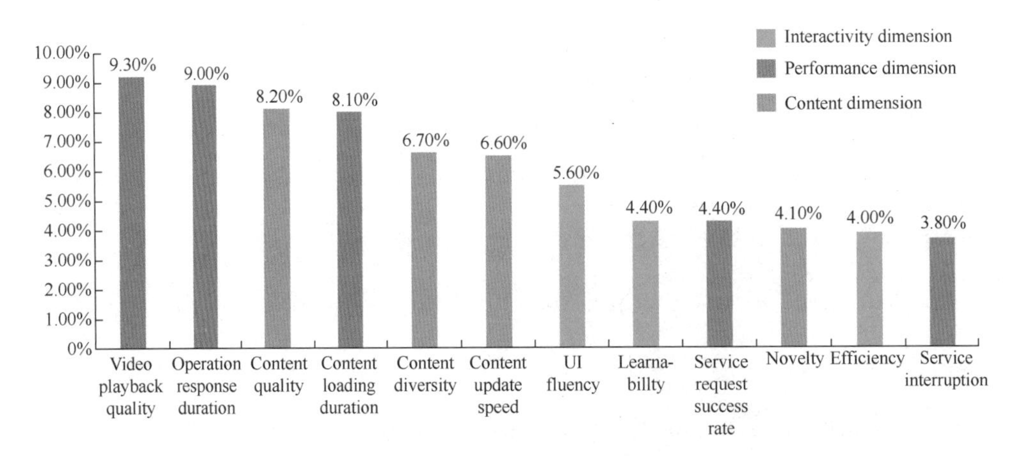

Source: China Video Service Users' Experience Standards Working Group

Figure 3-1 Influence of different factors on video experience

- Viewing experience: reflects freezing and artifact impairments and their impact on users' viewing experience.
- Interactive experience: reflects the response time after each operation, and can

be decomposed into initial loading duration and channel switchover duration.

- Video quality: reflects the quality of the content source and is usually measured by the content definition, such as UHD, HD, or SD.

The working group analyzed the influential factors of video experience and summed up three factors, in the *Video Experience Assessment Standard (1.0)* based on the interview result. namely content quality, interactive experience, and viewing experience (as shown in Figure 3-2).

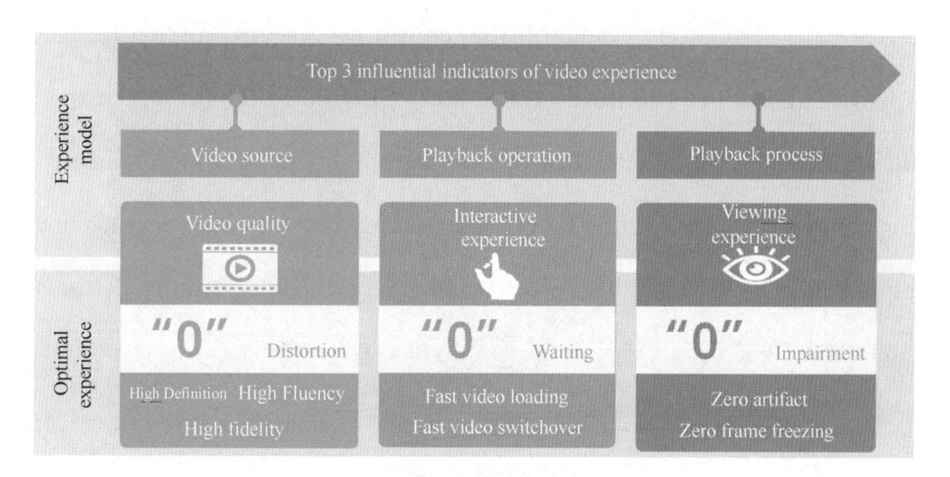

Figure 3-2 Main factors of video experience

3.1.1 Video Quality

The video quality factors are as follows.

- Video resolution: A higher video resolution indicates higher definition and higher video quality. For example, the resolution of a SD video is 720 × 576 and that of a FHD video is 1920 × 1080.
- Video bit rate: A higher video bit rate (such as 1 Mbit/s) indicates less

video source image loss, higher definition, and higher video quality.

- Encoding type: The video compression ratio varies according to the encoding type (like H.264 or H.265). Under the same bandwidth condition, advanced encoding technology provides higher video quality.
- Video frame rate: A higher video frame rate indicates smoother video playback and higher video quality. For example, the frame rate of a SD video is 24 FPS and that of a UHD video is 60 FPS.

Not only video source factors (the definition) but also influential factors in specific scenarios should be taken into consideration for video quality evaluation. For example, the size of the display device is a major influential factor (as illustrated in Figure 3-3). Larger screen requires higher video source resolution and bandwidth. Therefore, users' experience is usually better on a mobile phone than on a TV with the same video source.

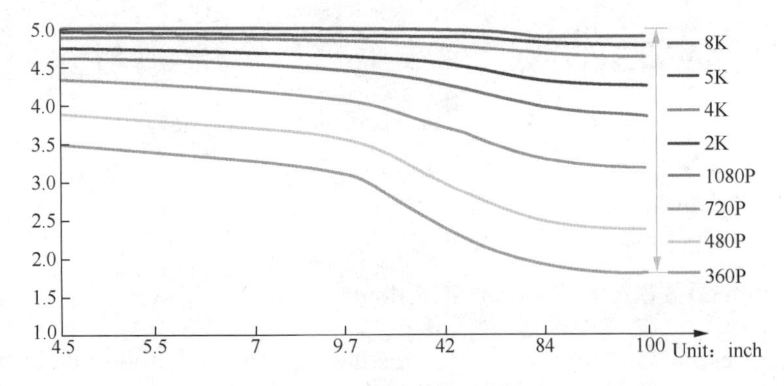

Figure 3-3 Impact of screen size on video quality

3.1.2 Interactive Experience

Interactive experience reflects the convenience and efficiency of video service operations. The influential factors include the operation success rate of live TV and VOD services along with interaction delay. The interactive experience of a user is mainly affected by the response time of the video system, including the response time of EPG operations, initial loading duration, channel zapping time, and response time of fast-forward operations. The interactive experience counters that concerned users may vary according to the video service type.

1. Interactive experience assessment standard for live TV services

Users who accustomed to traditional TVs have high expectations for channel switchover duration. Therefore, channel switchover duration is the most important factor that affects interactive experience. During live TV, longer channel switchover duration leads to poorer user experience. Table 3-1 lists the interactive experience assessment standard for live TV services according to the users' experience test and analysis of intensity response data of human eyes.

Table 3-1 Relationship between the interactive experience score and channel switchover duration of live TV services

Score	Channel Switchover Duration (Millisecond)	Comment
Excellent (4 ~5)	≤ 500	(4~5): 4<Interactive Experience Score ≤ 5
Good (3~4)	500~1000	(3~4): 3<Interactive Experience Score ≤ 4 500~1000: 500<Channel Switchover Duration ≤ 1000
Fair (2~3)	1000~2000	(2~3): 2<Interactive Experience Score ≤ 3 1000~2000: 1000<Channel Switchover Duration ≤ 2000
Poor (1~2)	2000~4000	(1~2): 1<Interactive Experience Score ≤ 2 2000~4000: 2000<Channel Switchover Duration ≤ 4000
Bad (1)	>4000	

2. Interactive experience assessment standard for VOD services

As the operation response of VOD services is always slower than live TV, users can bear longer interaction duration for VOD services. The two-second rule is used to measure the interaction duration in the industry. Most users find it acceptable if the initial video loading duration of a VOD service is within 2 s. If the duration exceeds 2 s, 10% more users will cancel the service with 1 s added. If the duration exceeds 10 s, most users will cancel the playback.

In addition, the working group found that users' interactive experience expectations are affected by usage habits and vary greatly according to the screen size. For example, users have higher requirements for initial video loading duration on TV than on mobile phone or on tablet, as shown in Figure 3-4. With the optimization of network transition, improvement of terminal performance, and increment in multi-screen video services, users will have higher interactive experience expectations and their expectations for overall interaction delay on different terminals will converge.

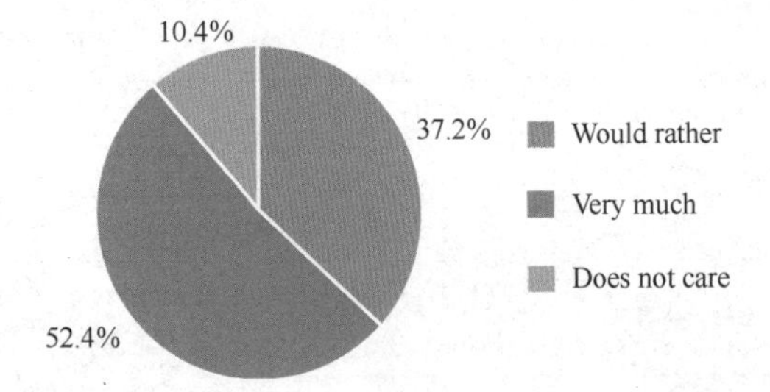

Source: China Video Service Users' Experience Standards Working Group

Figure 3-4 Users' expectations for consistent interaction delay on TVs and on mobile phones

Table 3-2 lists the interactive experience assessment standard for VOD services according to the users' experience test and analysis of intensity response data of human eyes.

Table 3-2 Relationship between the interactive experience score and initial video loading duration of VOD services

Score	Loading Duration on TVs (Millisecond)	Loading Duration on Mobile Phones (Millisecond)	Comment
Excellent (4~5)	≤ 1000	≤ 1000	(4~5): 4<Interactive Experience Score ≤ 5
Good (3~4)	1000~2000	1000~3000	(3~4): 3<Interactive Experience Score ≤ 4 1000~2000: 1000<Loading Duration ≤ 2000 1000~3000: 1000<Loading Duration ≤ 3000
Fair (2~3)	2000~5000	3000~5000	(2~3): 2<Interactive Experience Score ≤ 3 2000~5000: 2000<Loading Duration ≤ 5000 3000~5000: 3000<Loading Duration ≤ 5000
Poor (1~2)	5000~8000	5000~10000	(1~2): 1<Interactive Experience Score ≤ 2 5000~8000: 5000<Loading Duration ≤ 8000 5000~10000: 5000<Loading Duration ≤ 10000
Bad (1)	>8000	>10000	

3.1.3 Viewing Experience

Viewing experience reflects video quality deterioration such as image incoherence or image exception caused by artifacts, mosaic, frame freeze, and lip sync error during video playback. Possible causes include poor capabilities of network and service platforms as well as loss of synchronization between the platforms. For example, network packet loss may cause artifacts during live TV playback (as following Figure shows), packet delay due to network or platform cache may cause frame freezing during VOD playback (as illustrated in following figure).

In addition, the working group found that users' viewing experience expectations are affected by usage habits and vary greatly according to the screen size. For example, users have higher tolerance of frame freezing on mobile phone than on TV. As a result, frame freezing leads to poorer users' experience on TV but has less impact on users' experience on a mobile phone.

Figure Artifacts during live TV playback

Figure Frame freezing during VOD playback

3.1.4 Comprehensive Evaluation of Video Experience

Comprehensive video experience scores are calculated based on the preceding quantitative counters and are as close to actual user experience of video services as possible. The video quality, interactive experience and viewing experience factors are mainly used to measure video experience, as illustrated in Figure 3-5.

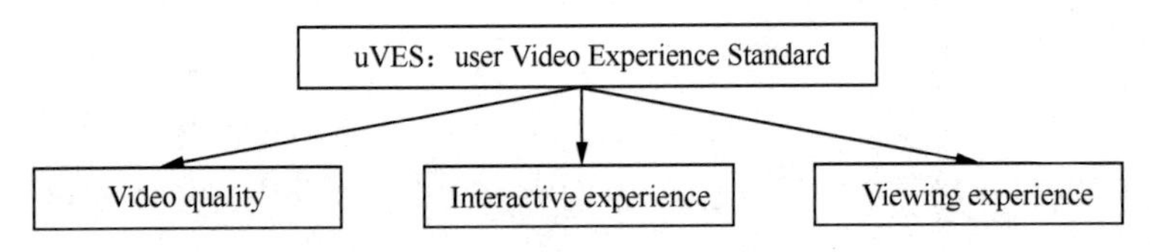

Figure 3-5 Counters of uVES scores

The *Video Experience Assessment Standard (1.0)* defines the counters in three aspects, provides the quantitative calculation method of comprehensive video experience scores, and defines the user Video Experience Standard (uVES) score. The uVES score reflects the users' video experience on a specified terminal at a specified time, see Table 3-3.

Table 3-3 Relationship between the uVES scores and users' experience

Score	Service Experience/User Satisfaction
Excellent (4~5)	Users' experience is excellent. Users' requirements are met or even exceeded
Good (3~4)	Users' experience is good. Users are willing to continue using the service
Fair (2~3)	Users' experience is fair. Most users can bear the service and will continue using it for a long period
Poor (1~2)	Users' experience is poor. Some users cancel the service or make complaints
Bad (1)	Users' experience is bad. Most users can't bear the service and cancel it

- The uVES score is a quantitative method used by the industry to get a result that is as close to the actual result as possible. Compared with the mean opinion score (MOS) calculated in the lab environment, the uVES score is less accurate but can be used to perform real-time online evaluation under the live network.

- In contrast to the previous calculation methods used to evaluate the video service quality that only counters at the network level, the uVES score is calculated based on counters at both service and network levels.
- The counters for calculating the uVES score may vary according to service and scenario. The application scope of the uVES score is defined in the standard and can be modified based on the site requirements.

3.2 The First National Evaluation of Users' Experience Based on the Publication of the Standard

The publication of the standard lays the foundation for real-time online evaluation of video experience based on networks, terminals, and scenarios. To test the practicability of the standard, major telecom carriers and Internet video CPs in China organized the first large-scale users' experience evaluation activity after the publication of the *Video Experience Assessment Standard (1.0)*.

Currently, video services in China mainly include IPTV, Internet video and cable TV services. Only the sample data of IPTV services and Internet video services in some regions were collected for users' experience evaluation due to time restrictions. Sample data of cable TV services are still being collected and will be released in subsequent reports.

Some manufacturers in China have been developing and deploying tools for evaluating video experience. Only evaluation tools that pass the standard consistency test can be deployed at large scales. In the evaluation activity, users in some regions where the services were mature and the evaluation tools had passed the standard consistency test were selected as evaluation samples. At present, multiple telecom carriers, cable TV carriers, and Internet video CPs are deploying the users' experience evaluation system. With the implementation of related work, sample data of users' experience on terminals of all mainstream video CPs will be collected.

3.2.1 Evaluation of IPTV Services

In this activity, evaluation tools were deployed on IPTV terminals to collect data of the VOD and live TV users of IPTV services provided by some provincial carriers in China Telecom and China Unicom.

- The definitions of the evaluation samples include PAL (720 × 576), 720P (1280 × 720), 1080P (1920 × 1080), and 4K (3840 × 2160). The screen sizes of users' TVs are mainly 42 inches.

- The consecutive real-time data of a week in October, 2016 were used as the sample data. The evaluation software collected data from users' IPTV STBs every five minutes. About one million samples were collected.

3.2.2 Evaluation of the Internet Video Services

In this activity, the services of some mainstream Internet video CPs in China, including Youku, iQIYI, Tencent, LETV, Mango TV, FengXing TV, were evaluated. The evaluation tools released by third-party video experience evaluation companies (Beijing BonRee Co., Ltd. and Tingyun Technology Co., Ltd.) were used, and some friendly users were invited to help with the evaluation.

- Content samples used for the evaluation include popular movies, TV series, and variety shows. The screen sizes of users' PCs and mobile phones are mainly 19 inches and 5.5 inches respectively.
- The definitions of the evaluation samples include 270P (480 × 272 or 640 × 272), 360P (640 × 352, 640 × 360, or 960 × 540), 720P (1280 × 720), 1080P (1920 × 1080), and 4K (3840 × 2160).
- The consecutive real-time data of a week in October, 2016 were used as the sample data. The evaluation software collected data from users' terminals every five minutes. About five million samples were collected.
- The evaluation covered first-tier cities (Beijing, Shanghai, Guangzhou, and Shenzhen), second-tier cities (Hangzhou, Chengdu, Xi'an, Shenyang), and

third-tier cities (Haikou and Nanning).

- The broadband networks and mobile networks for evaluation were provided by China Telecom, China Mobile, China Unicom, Great Wall Broadband Network (GWBN), China Education and Research Network (CERNET), along with China Tietong.

4. Video Experience in China Varies Greatly and has Much Room for Improvement

After analyzing the collected sample data, the working group found that the video experience varies greatly according to the service type and CP. For video services guaranteed by specified networks, the viewing experience was good but the interactivity performance needed to be improved. For video services transmitted on the public Internet, users' experience was greatly affected by network traffic and needed to be improved through technological measures.

4.1 Overall Users' Experience of IPTV Services Is Higher Than That of the Internet Video Services

According to the sample data, the overall users' experience of IPTV services is guaranteed and that of the Internet video services is greatly affected by networks. The average uVES score of IPTV sample users is 3.06 (good). The average uVES score of the Internet video services on PCs is 1.79 (poor) and that on mobile phones is 2.53 (Fair). Compared with IPTV services, the Internet video services have much room for improvement. The IPTV services are transmitted through private networks and the network transmission quality is guaranteed. The Internet video services are transmitted through public networks and need to share network resources with other Internet services. Therefore, users' experience of the Internet video services can't be guaranteed during busy hours.

The reason why the uVES score of the Internet video services is lower than that of IPTV services is that the video quality of the Internet video services is greatly affected by the network and falls significantly behind that of IPTV services. The video quality score of the Internet videos on PCs is 1.98 and that on mobile phones is 2.74,due to different screen size, as shown in Figure 4-1.

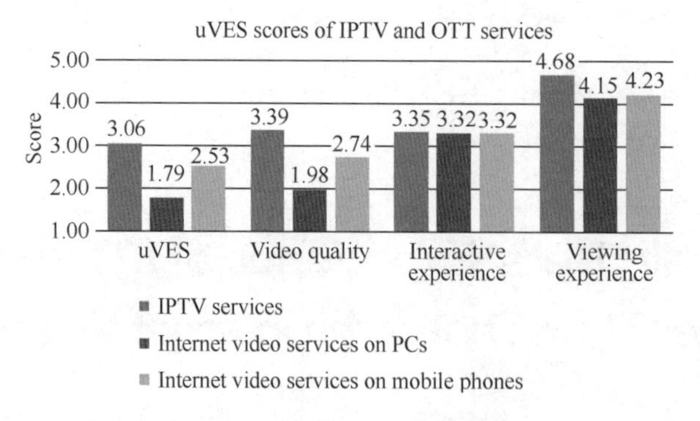

Source: China Video Service Users' Experience Standards Working Group

Figure 4-1 uVES scores of IPTV services, Internet video services on PCs, and the Internet video services on mobile phones

4.2 Good Users' Experience of IPTV Services Provided by Telecom Carriers

In this activity, users' experience of IPTV live services and IPTV VOD services provided by some provincial telecom carriers in China on TVs (mainly 42-inch screens) was evaluated.

4.2.1 Overall Evaluation of Users' Experience of IPTV Services Provided by Telecom Carriers

1. The overall users' experience of IPTV services provided by telecom carries is good. The viewing experience behaves the best but the video quality and interactive experience have much room for improvement as Figure 4-2 reveals

- The average uVES score is 3.06 (good) and the average score of each factor is above 3.
- The average score of viewing experience is 4.68 (excellent). About 8‰ of users encounter frame freezing or artifacts during video playback.
- Interactive experience is a weakness of IPTV services and has much room to improve. The average score is 3.35 and the minimum score is 1.
- Video quality has much room for improvement.

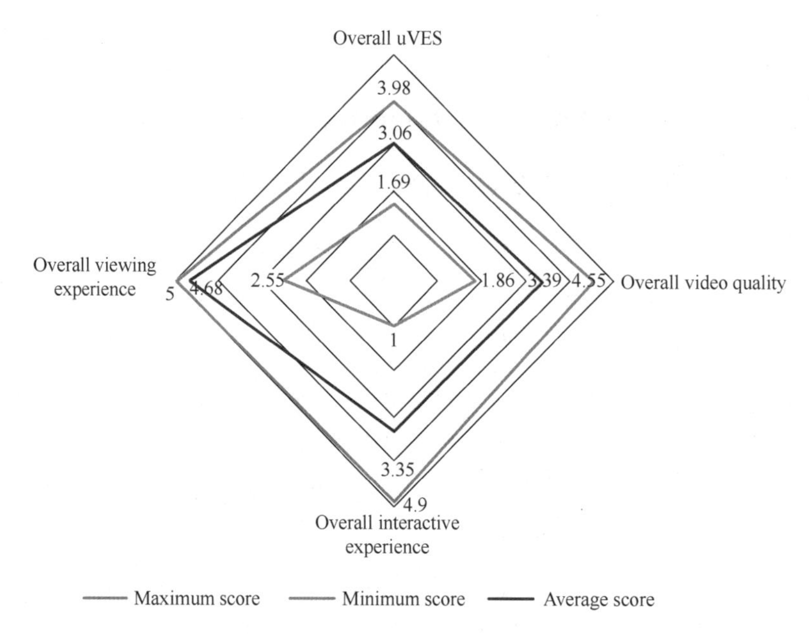

Source: China Video Service Users' Experience Standards Working Group

Figure 4-2 Overall evaluation of IPTV services provided by carriers

2. The average uVES score of live TV services is above 3 (good), which shows the network advantage, and Figure 4-3 shows the details

- The average score of each factor for live TV services is above 3.
- The viewing experience score is 4.72 (excellent).
- The average score of interactive experience is 3.35, which is the lowest and has much room for improvement (Users have high requirements for live TV channel switchover. The dedicated IPTV service system needs to be upgraded.)

3. The average uVES score of VOD services is above 3 (good), and Figure 4-4 indicates the detailed situation

- The average score of each factor for VOD services is above 3 (good).

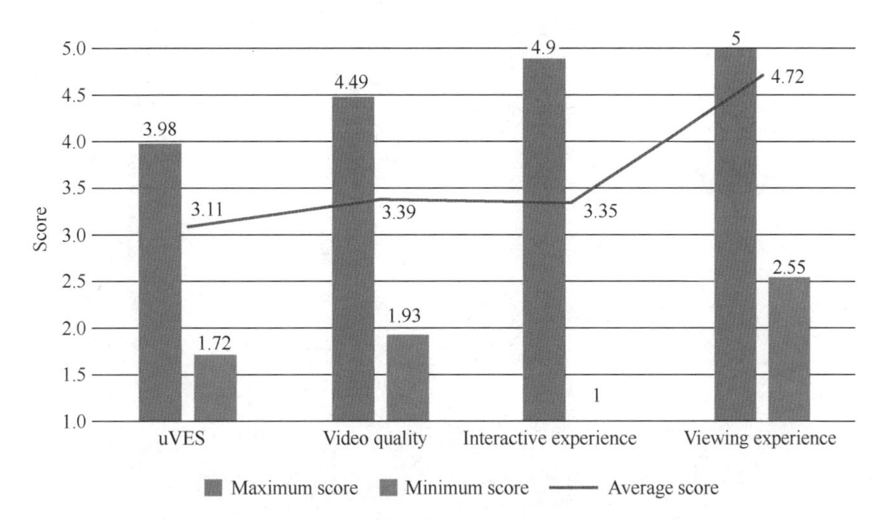

Source: China Video Service Users' Experience Standards Working Group

Figure 4-3 Users' experience analysis of IPTV live services

- Viewing experience is excellent and the score is almost 5, indicating that quality can be guaranteed by private networks.
- Video quality of VOD services gets the lowest score and needs to be improved.
- The average score of interactive experience is 3.7 and has room for improvement.

4. The interactive experience and viewing experience of 4K VOD services can be improved

According to the evaluation result, the video quality score of 4K VOD services is higher than that of HD VOD services, which shows the advantage of 4K technology, as shown in Figure 4-5.

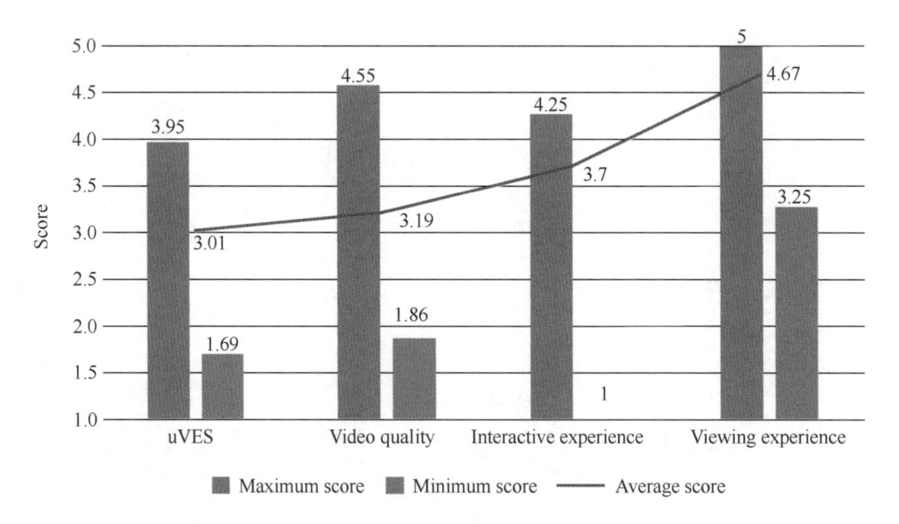

Source: China Video Service Users' Experience Standards Working Group

Figure 4-4 Users' experience analysis of IPTV VOD services

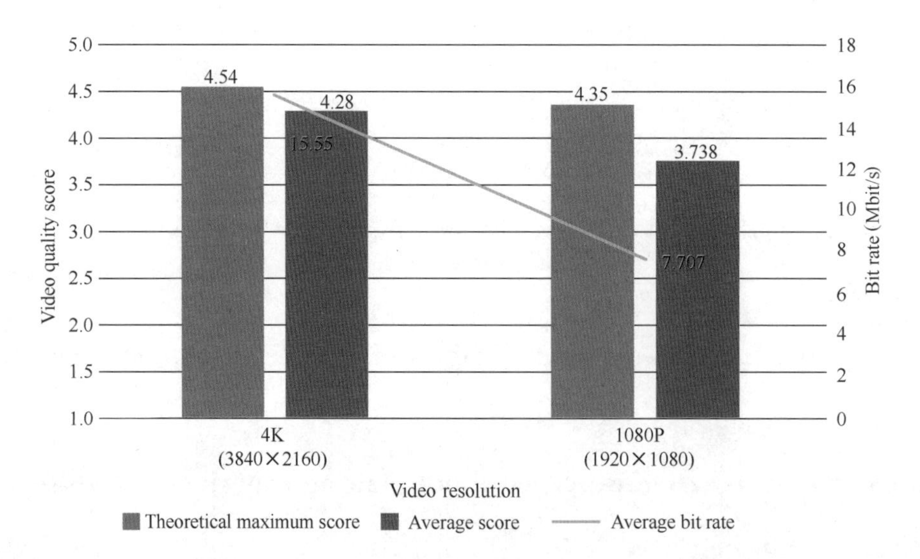

Source: China Video Service Users' Experience Standards Working Group

Figure 4-5 Video quality scores of IPTV VOD services with various resolutions

However, the interactive experience score of 4K VOD services is low (Figure 4-6 shows) because the average 4K bit rate (15 Mbit/s) is higher than the average HD (1080P) bit rate (7.707 Mbit/s). Currently, carriers generally use HTTP or HLS protocol for 4K IPTV VOD services. The loading duration of 4K IPTV VOD content is close to 3s, which is longer than the average loading duration (1.279s) of HD or SD videos over RTP or RTSP protocol. When the video loading duration exceeds 2s, the services may lose attraction and lead to customer churn.

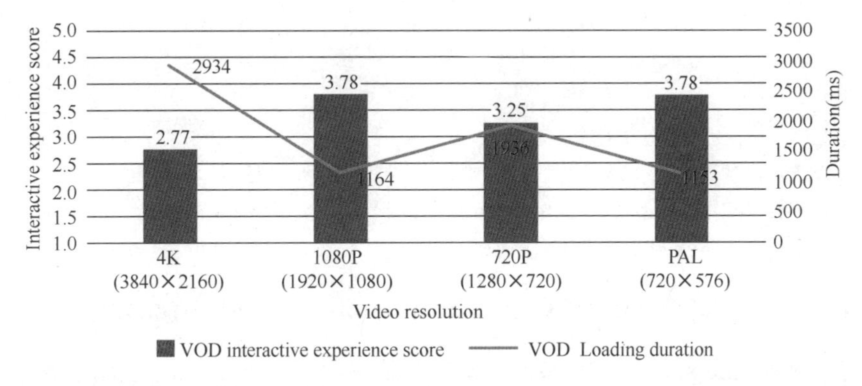

Source: China Video Service Users' Experience Standards Working Group

Figure 4-6 Interactive experience scores of IPTV VOD services with various resolutions

Frame freezing occurs more often during 4K VOD playback than HD or SD, as indicated by Figure 4-7. As 4K content increases in IPTV services, the network transmission capability needs to be improved and more powerful terminals need to be deployed.

4.2.2 Great Video Quality of IPTV Services Provided by Telecom Carriers

The working group evaluated the video quality of IPTV services provided by some provincial telecom carriers in China on TVs (mainly 42-inch screens).

The evaluation results are as follows.

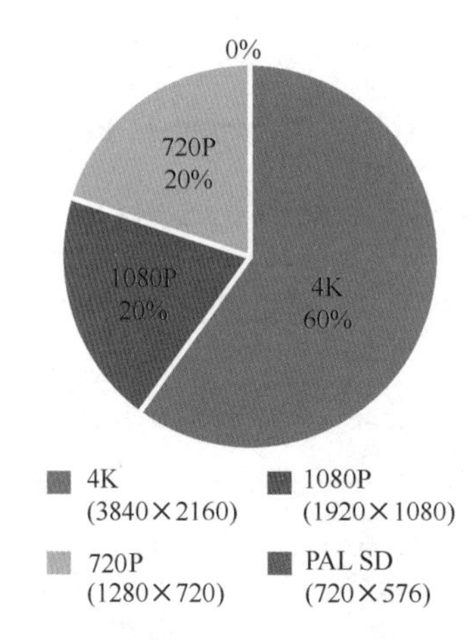

Source: China Video Service Users' Experience Standards Working Group

Figure 4-7 Distribution of frame freezing occurrence probabilities during the playback of IPTV VOD content with various resolutions

1. The average score of IPTV video quality exceeds 3 (good)

The average score of video quality of IPTV services is 3.39 (good), and Figure 4-8 provides the details. The bit rates configured for videos with various resolutions are appropriate and obey the following rule: The video quality score increases with the video resolution and bit rate. The average quality scores of 4K videos and PAL SD videos are 4.28 (highest) and 2.763 (lowest) respectively. The video quality score of 1080P videos is close to that of 720P videos.

2. HD content becomes mainstream for IPTV

According to the data, 1080P (1920 × 1080) content has the highest viewership and occupies more than 70% watching records, indicating that HD content has become mainstream. Most content resolutions are standard SD/720P/1080P/4K, the non-standard resolutions (including 544 × 578 and 528 × 576) is only 1.68%, and details are shown in Figure 4-9.

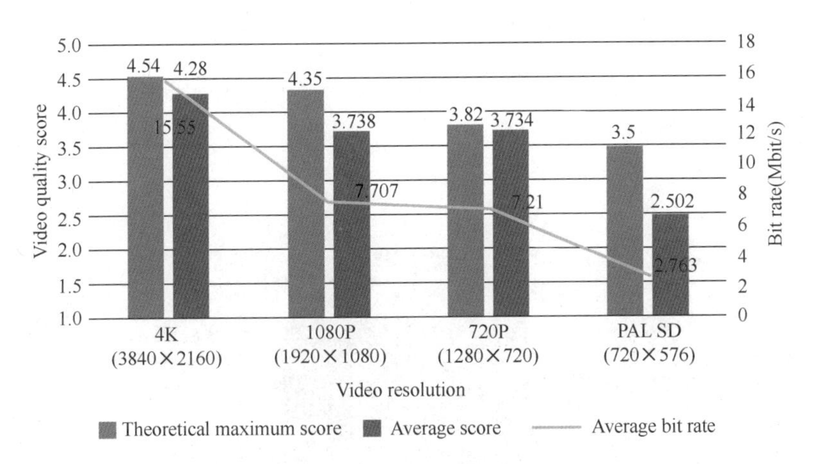

Source: China Video Service Users' Experience Standards Working Group

Figure 4-8 Video quality scores of IPTV services with various resolutions

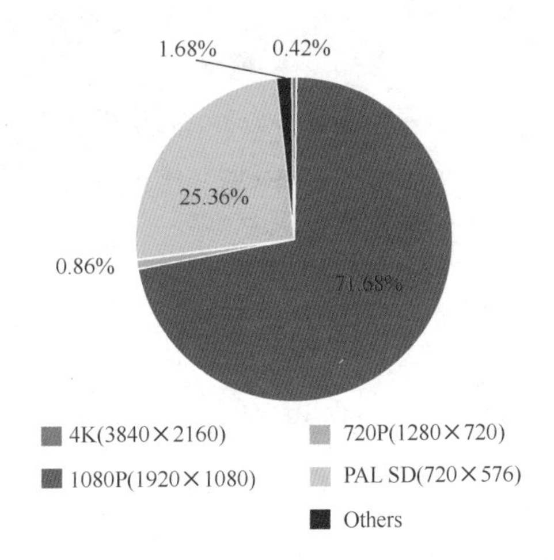

Source: China Video Service Users' Experience Standards Working Group

Figure 4-9 Distribution of watched videos with various resolutions of IPTV services

3. The video quality of IPTV live services is higher than that of IPTV VOD services

- Among IPTV live TV content with various resolutions, as illustrated in Figure 4-10, the viewership of HD content is 76.05%, which is the highest. The viewership of HD live TV content is 26.1%, higher than that of HD VOD content.

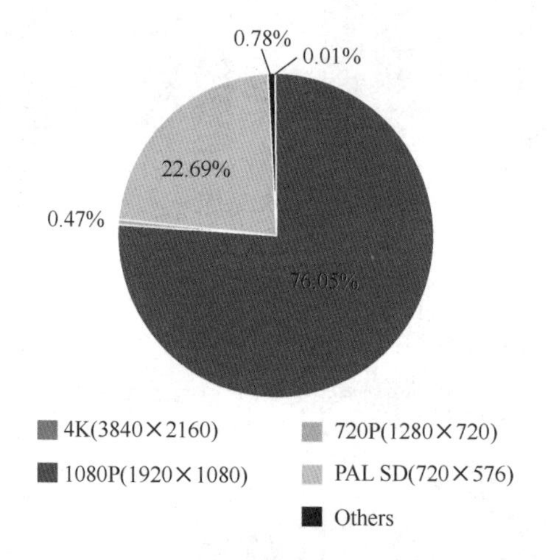

Source: China Video Service Users' Experience Standards Working Group

Figure 4-10 Distribution of resolutions of IPTV live TV content

- SD content occupies a higher percentage in IPTV VOD services than in IPTV live services due to the media sources provided by carriers, that lowers the average quality score of VOD services. HD content will occupy a higher percentage in VOD services as time goes by.
- Currently, 4K content is insufficient in IPTV live services and occupies a percentage of 2.53% in IPTV VOD services in China, as shown in Figure 4-11.

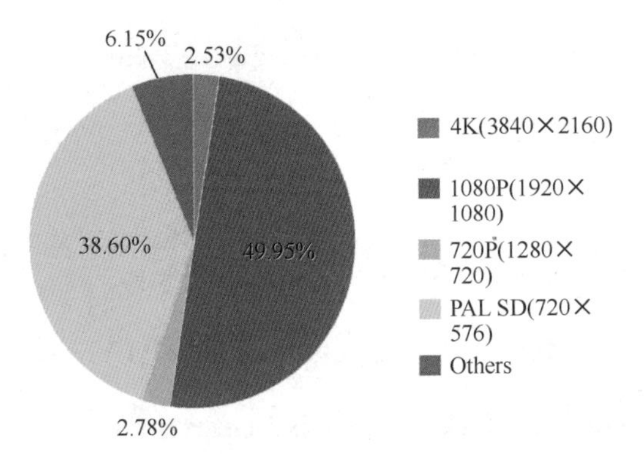

Source: China Video Service Users' Experience Standards Working Group

Figure 4-11 Distribution of resolutions of IPTV VOD content

- Content with a resolution of 720P or higher occupies a percentage of 55.26% in IPTV services, indicating that the quality of IPTV video resources is improving and users are becoming more accustomed to watching HD content.

4.2.3 Good Interactive Experience of IPTV Services Provided by Telecom Carriers

Among the interactive experience counters of IPTV services, performance of the response speed of menu operations and operation success rate is excellent. Only the counters including initial video loading duration of VOD services, channel switchover duration of live TV services, and response speed of fast-forward and rewind operations require special attention. For details, see Table 3-1 and Table 3-2. Users who are accustomed to traditional TV services and Internet VOD services have different interactive experience expectations for live TV services and VOD services. Compared with initial loading duration of VOD services, users have higher expectations for channel switchover duration of live TV services.

The evaluation shows the following conclusions.

1. The overall interactive experience of IPTV services is good. The interactive experience of VOD services is better than that of live TV services, as Figure 4-12 shown

- The average scores of overall interactive experience is above 3. The average scores of interactive experience of both VOD services and live TV services are above 3. Because users' expectations for VOD services are lower, the interactive experience score of VOD services is higher than that of live TV services.

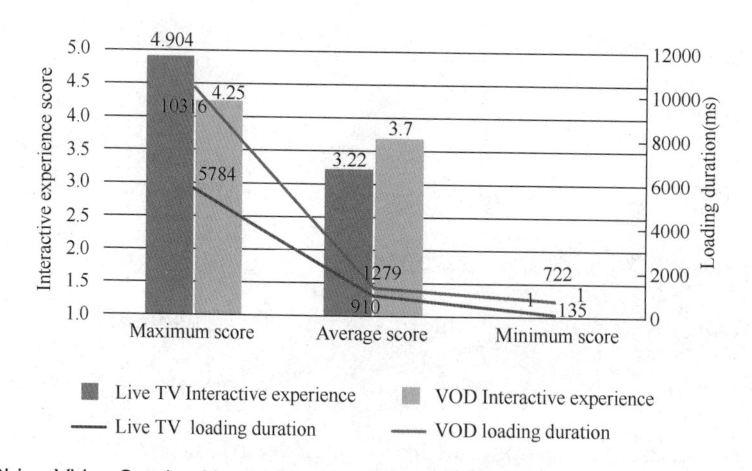

Source: China Video Service Users' Experience Standards Working Group

Figure 4-12 Interactive experience scores of IPTV VOD services and IPTV live services

- The average initial video loading duration of VOD services is 1279 ms, the maximum duration is 10316 ms, and the minimum duration is 722 ms. The average channel switchover duration of live TV services is 910 ms, the maximum duration is 5784 ms, and the minimum duration is 135 ms. The initial video loading duration of 23.09% VOD samples is within 1s, and the channel switchover duration of 56.12% live TV samples is within 1s. Both counters are variable.

2. The interactive experience of IPTV live services is good. The channel switchover duration is generally within 1s

The average interaction duration of sample data in live TV services is 910 ms. 50.05% of users' experience scores are higher than the average score, 56.12% of users' channel switchover duration is within 1s, 93.06% of users' channel switchover duration is within 2s, and 0.65% of users' channel switchover duration is more than 4s. The channel switchover duration has room for improvement, and Figure 4-13 illustrates the details.

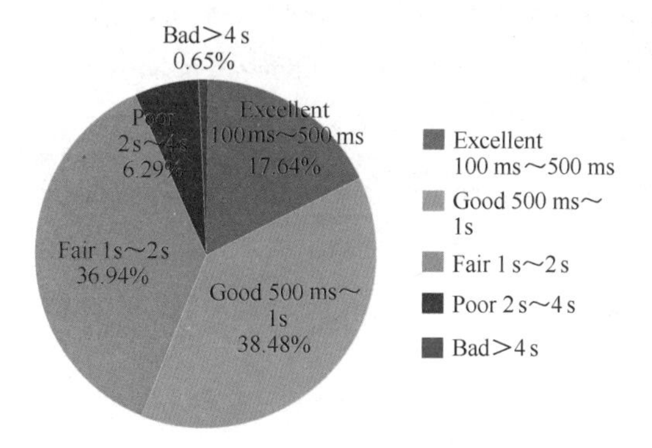

Source: China Video Service Users' Experience Standards Working Group

Figure 4-13 Distribution of channel switchover durations of IPTV live services

3. The interactive experience of IPTV VOD services is good. The average initial video loading duration is within 2s

The average initial video loading duration of VOD services is 1279 ms. Over all the sample data, 47.6% of users' interactive experience scores are higher than the average score, 92.9% of users' initial video loading duration is within 2 s, almost 70% of users' initial video loading duration is between 1s and 2 s, 23.09% of users' initial video loading duration is within 1s, and 0.01% of users' initial video loading duration is longer than 8s and Figure 4-14 shows the details.

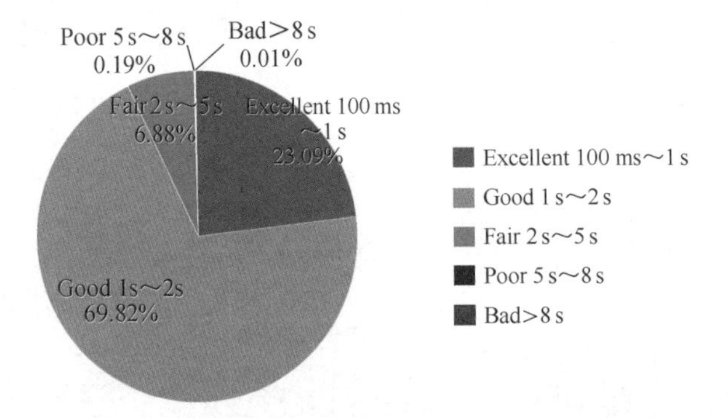

Source: China Video Service Users' Experience Standards Working Group

Figure 4-14 Distribution of initial video loading durations of IPTV VOD services

4.2.4 Excellent Viewing Experience of IPTV Services Provided by Telecom Carriers

1. The viewing experience of IPTV services is excellent. Users seldom encounter artifacts or frame freezing

The average score of viewing experience of IPTV services is almost 5. Only 0.1‰ of IPTV users encounter frame freezing or artifacts during video playback. The viewing experience of live TV playback is better than that of VOD playback. The frequency of artifact occurrence during live TV playback is 0.0166‰ and that of frame freezing occurrence during VOD playback is 0.2‰ .

2. Frame freezing occurs more frequently during 4K VOD playback than lower resolution

The number of frame freezing times during 4K VOD playback accounts for 60% of the total number of frame freezing times, because 4K content requires higher bandwidth for transmission, as illustrated in Figure 4-15.

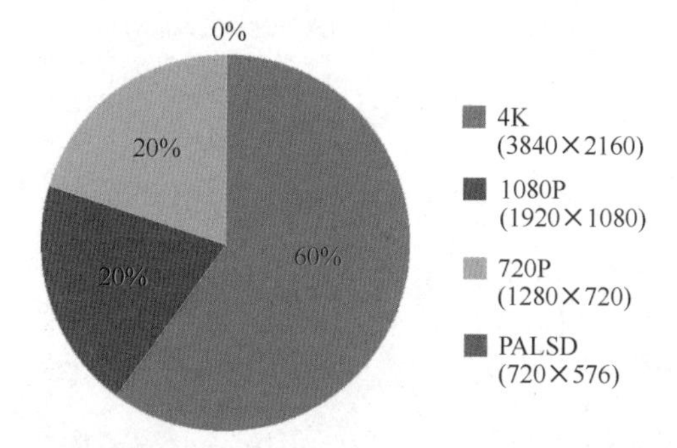

Source: China Video Service Users' Experience Standards Working Group

Figure 4-15 Distribution of frame freezing occurrence probabilities during the playback of IPTV VOD content with various resolutions

4.3 Insufficient Assurance for Users' Experience of Internet Video Services

In this activity, the Internet video VOD services of some provinces in

China were used for users' experience evaluation. Content with the same resolution and bit rate was selected for evaluation of users' experience on PCs (mainly 19-inch screens) and mobile phones (mainly 5.5-inch screens).

4.3.1 Overall Evaluation of Users' Experience of Internet Video Services

1. Users' experience of Internet video services varies from different terminals and the video quality and interactive experience have much room for improvement.

- The average uVES score of users' experience on PCs is 1.79 and 82.61% of users' experience is poor (uVES score ranging from 1 to 2).
- The average uVES score of users' experience on mobile phones is 2.53, which is higher than that on PCs. Among the sample data, 5.63% of users' experience is good and 0.48% of users' experience is excellent.
- The average score of the video quality on PCs is 1.98, which is much different from that on mobile phones.
- From the interactive experience and viewing experience perspectives, the score of users' experience on PCs is close to that on mobile phones, while Figure 4-16 illustrates the details.

2. There is no significant difference among users' experience of Internet video services in different regions on the same tier, and Figure 4-17 shows the evidence

- The analysis's result of experience data obtained from sample data in Beijing, Shanghai, Guangzhou, Shenzhen (first-tier), Hangzhou, Shenyang,

Xi'an, Chengdu (second-tier), Haikou, Nanning (third-tier), indicates that the user experience scores of Internet video services in cities on the same tier are not quite different. The average score of users' experience on PCs and that on mobile phones are both close to the average score obtained in the evaluation.

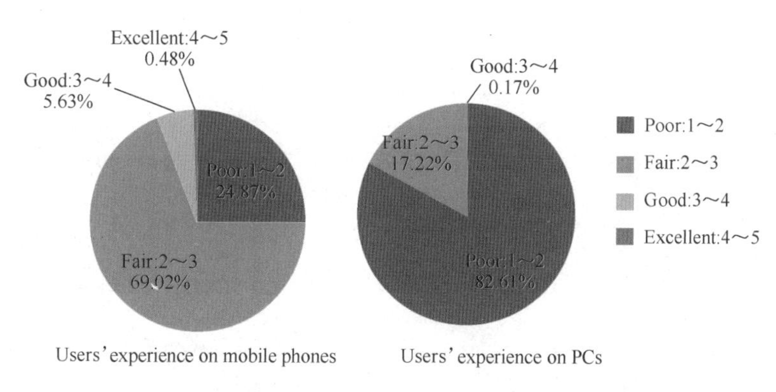

Source: China Video Service Users' Experience Standards Working Group

Figure 4-16 Distribution of users' experience scores of Internet video services

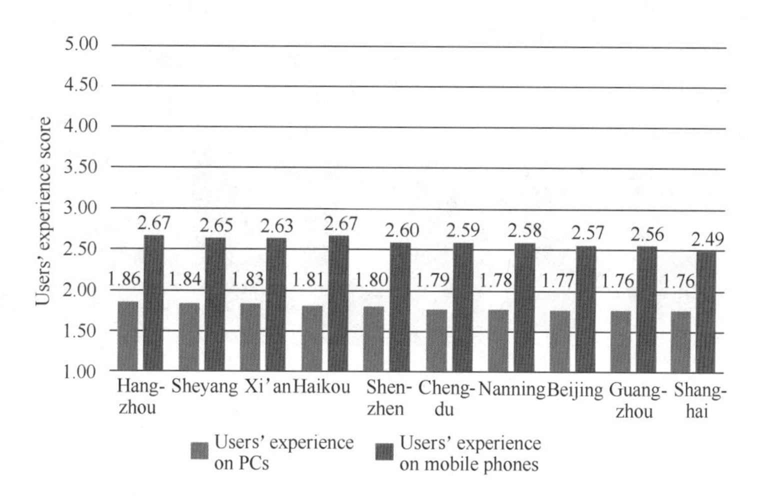

Source: China Video Service Users' Experience Standards Working Group

Figure 4-17 uVES scores of user experience of Internet video services in different cities

- Among the cities involved in the evaluation, Hangzhou gains the highest uVES score. The average score of users' experience on PCs is 1.86 and that on mobile phones is 2.67.
- The uVES score of user experience in Shanghai is the lowest due to the high population pressure. The average score of users' experience on PCs is 1.76 and that on mobile phones is 2.49, which are similar to other cities.

3. There are significant differences in users' experience of services provided by different Internet video CPs

In this activity, services provided by seven main video CPs in China were evaluated. The evaluation result indicates that users' experience scores of Internet video services are low due to lack of coordination with carriers' networks, and details can be get from Figure 4-18 and Figure 4-19.

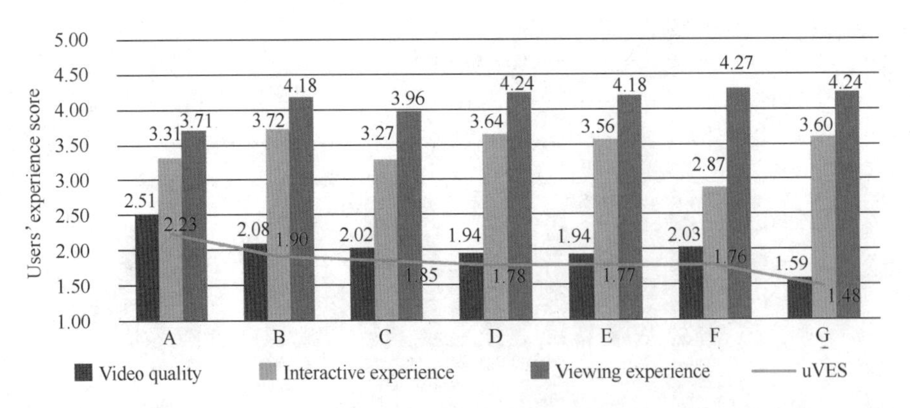

Source: China Video Service Users' Experience Standards Working Group

Figure 4-18 Users' experience scores of Internet video services provided by different CPs (on PCs)

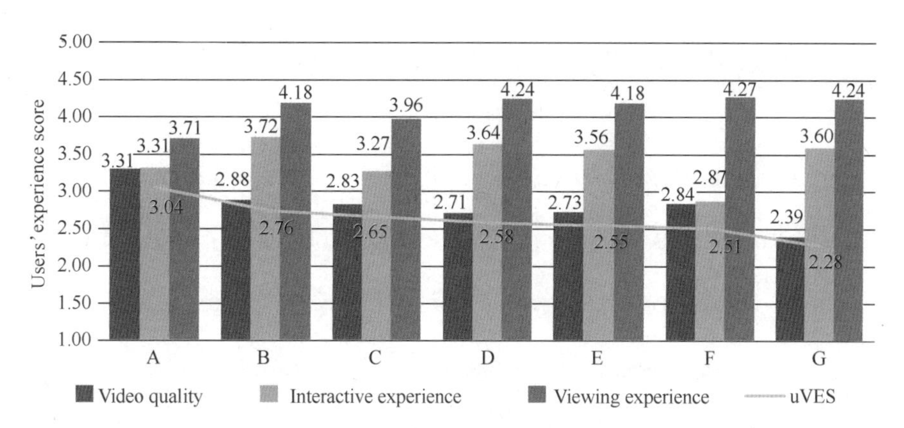

Source: China Video Service Users' Experience Standards Working Group

Figure 4-19 Users' experience scores of Internet video services provided by different CPs (on mobile phones)

- The uVES score of users' experience of services provided by CP-A is the highest. The uVES score of users' experience on PCs is 2.23 and that on mobile phones is 3.04. The interactive experience and viewing experience scores of CP-A are slightly lower than those of other CPs. The advantage of CP-A is that its video quality is higher than other CPs. The score of CP-A's video quality on PCs is 2.51 and that on mobile phones is 3.31, which are respectively higher than those of CP-B (ranked secondly) by over 0.4. This is the key why CP-A's uVES score of user experience is the highest.
- The uVES score of users' experience of services provided by CP-G is the lowest. The uVES score of users' experience on PCs is 1.48 and that on mobile phones is 2.28. The interactive experience and viewing experience scores of CP-G are not much different from other CPs. However, the score

of CP-G's video quality is quite low. The score of CP-G's video quality on PCs is 1.59, and that on mobile phones is 2.39.

- The interactive experience score of CP-F is 2.87, which is lower than those of other CPs. However, the viewing experience score of CP-F is the highest. To ensure the playback smoothness, CP-F increases the initial VOD loading duration.
- The uVES scores of users' experience of services provided by CPs-B, C, D, and E are at the same level and in the middle range among those of the seven CPs. The interactive experience, viewing experience, and video quality scores of CPs-B, C, D, and E are relatively balanced. Internet video CPs mainly take the following measures to improve users' experience of video services.
- Increase the video bit rate to improve the video experience. However, high bandwidth is unavailable to support the high bit rate for Internet video services. As a result, frame freezing occurs more often after bit rate increases, and the viewing experience score lowers.
- Increase the initial video loading duration to buffer more data so that the video playback smoothness can be ensured. In this way, the viewing experience can be ensured to a certain extent, but the increase of the initial VOD loading duration affects the interactive experience. Therefore, the uVES score of users' experience does not increase.
- Keep video quality, interactive experience, and viewing experience in

balance. For example, ensure the video playback smoothness under the condition that the initial video loading duration is not long. To achieve this goal, video CPs must take multiple technical measures such as CDN sinking to become closer to users, new video coding technologies, and optimization of terminals' playback performance.

4. Significant differences exist in network carriers' ability to guarantee video experience

In this activity, users' experience of Internet video services provided by networks of China Telecom, China Mobile, China Unicom, GWBN, CERNET, and China Tietong was evaluated and compared. The evaluation and comparison results show that the carriers' networks have big differences in ensuring service quality, as shown in Figure 4-20 and Figure 4-21.

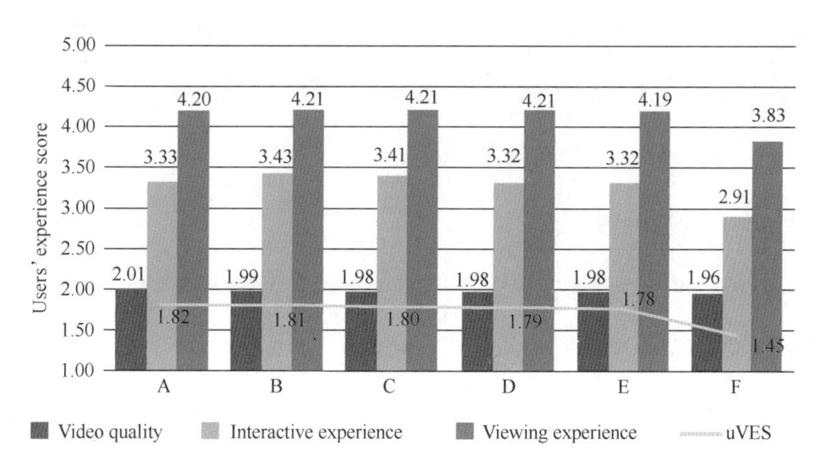

Source: China Video Service Users' Experience Standards Working Group

Figure 4-20 User experience scores of Internet video services from different carriers' networks (on PCs)

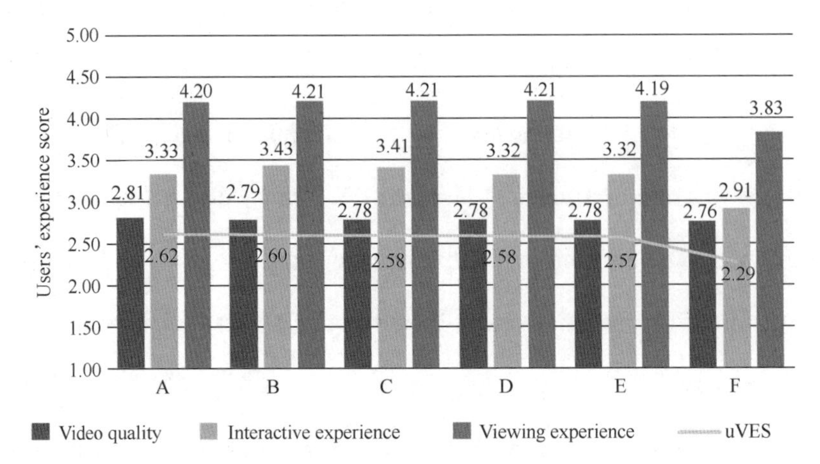

Source: China Video Service Users' Experience Standards Working Group

Figure 4-21 Users' experience scores of Internet video services from different carriers' networks (on mobile phones)

- The uVES score of carrier F is 1.45, which is far lower than the average uVES score (1.79) among these six carriers. The interactive experience and viewing experience scores of carrier F are respectively 2.91 and 3.83, which are lower than those of other carriers. The video quality score of carrier F is close to those of other carriers. The main problem lies in E2E network performance.
- The video quality, interactive experience, and viewing experience scores of carriers A, B, C, D, and E are similar, and therefore the uVES scores of these carriers are close. The network upgrading done by carriers in China to meet video service developing requirements has achieved some achievements.

4.3.2 Evaluation of the Quality of Internet Video Services

1. The overall quality of Internet video services is poor, and the video quality on mobile phones is better than that on PCs

- Mobile phones' screens are smaller. Therefore, for video streams with the same resolution and bit rate, the quality of mobile phones is better than that of PCs. The average score of the video quality on mobile phones is 2.74 and that on PCs is 1.98. For details, see figure 4-22.

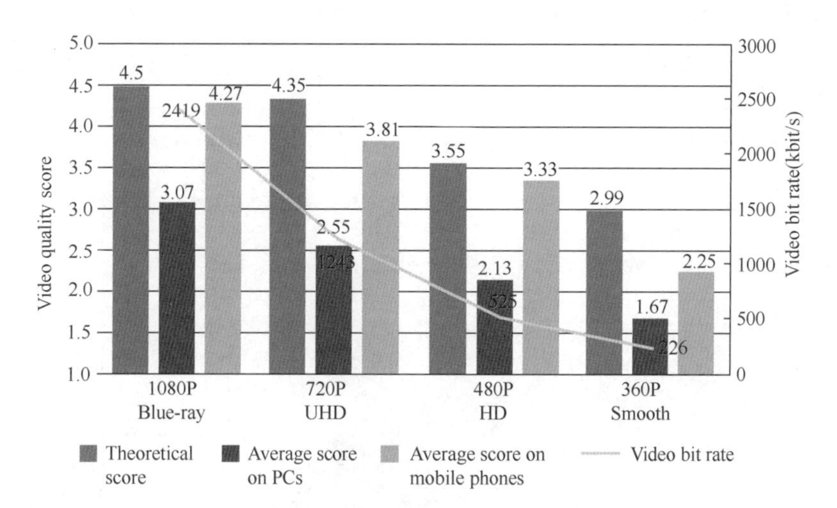

Source: China Video Service Users' Experience Standards Working Group

Figure 4-22 Quality scores of Internet video services with various resolutions

- The bit rate configured for Internet video services is appropriate and obeys the following rule. The video quality score increases with the video resolution and bit rate. For example, on a 5.5-inch mobile phone, the average score of the quality of 1080P videos is 3.07, which is the highest, and that of 360P videos is 1.67, which is the lowest.

- The video quality on PCs is generally lower than that on mobile phones. The average scores of the quality of 1080P and 360P videos on PCs are respectively lower than those on mobile phones by 1.2 and 0.62.
- The actual scores of the quality of Internet videos with various resolutions are quite different from the theoretical limits. The main reason is that the network bandwidth can't be ensured and the bit rate applied to Internet video services is low. Therefore, to improve the video quality, the bit rate must be increased.

2. Diversified resolutions are used in Internet video services

The sizes of PC screens and mobile phone screens are diversified. To ensure better users' experience, video CPs provide media sources with various resolutions for each video content. Therefore, the resolutions used in Internet video services are diversified, as illustrated in Figure 4-23.

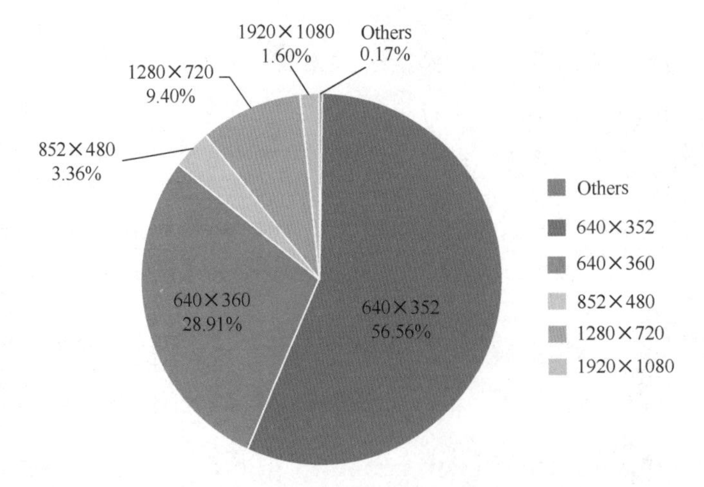

Source: China Video Service Users' Experience Standards Working Group

Figure 4-23 Distribution of resolutions of media sources in Internet video services

Currently, the same content of an Internet video service may use up to 10 resolutions, such as 270P, 360P, 480P, 720P, and 1080P. According to the analysis result of all mobile users' (mainly using 5.5-inch mobile phones) viewing records, the number of records for 640 × 352 and 640 × 360 content accounts for relatively large proportions, which are respectively 57% and 29%. Of all Internet video viewing records, the number of Non HD content (lower than 720P) records occupies over 80% , and the number of records for 720P and 1080P content occupies only 10%, as shown in Figure 4-24.

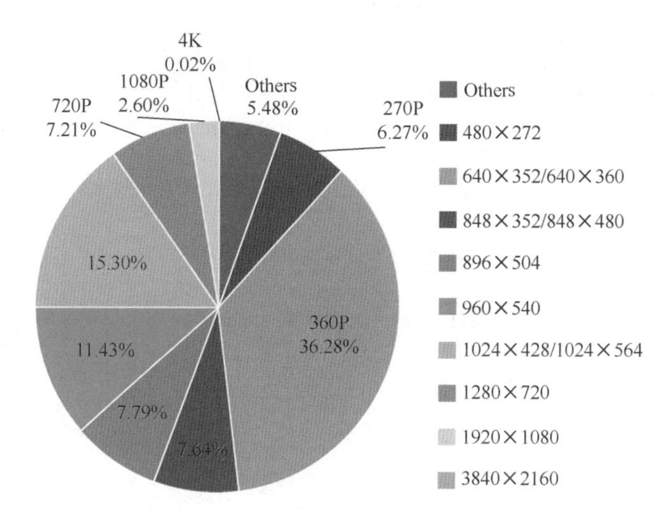

Source: China Video Service Users' Experience Standards Working Group

Figure 4-24 Distribution of resolutions of watched Internet videos

With the enhanced awareness of copyright in China, Internet video websites tend to adopt the "pay to watch" mode. As the number of paying users increases, the video experience requirements become higher and higher, and demand for videos with high resolutions (720P, 1080P, 2K, and 4K) will become the main stream.

4.3.3 Evaluation of Interactive Experience of Internet Video Services

The evaluation result of interactive experience of Internet video services is as follows.

1. The overall interactive experience of Internet video services is good

Table 4-1 shows that the average score of interactive experience of Internet video services is 3.32, the highest score is 4.68, and the lowest score is 1. Among the main factors, the average duration of initial video loading is 1660 ms and the maximum loading duration is 20310 ms. 35% of users' video loading duration is within 1 second (excellent), and 39% of users' video loading duration is between 1000 and 2000 ms (good). Only 6% of users' video loading duration exceeds 5s. The overall interactive experience of Internet video services is acceptable, as illustrated in Figure 4-25.

Table 4-1 Evaluation result of interactive experience of Internet video services

	Loading Duration (Millisecond)	Interactive experience Score
Maximum value	20310	1
Average value	1660	3.32
Minimum value	300	4.68

2. The interactive experience scores of different Internet video CPs are quite different

Among the video CPs involved in the evaluation, CPs-E, G, D, and B gain higher scores, and the interactive experience scores of these CPs are higher than the average interactive experience score. The interactive experience score of CP-B is the highest, with a initial video loading duration of 1076 ms. The interactive experience scores of CPs-F, C, and A are lower than the average interactive experience score and the longest duration of initial video loading is 3041 ms, and see Figure 4-26.

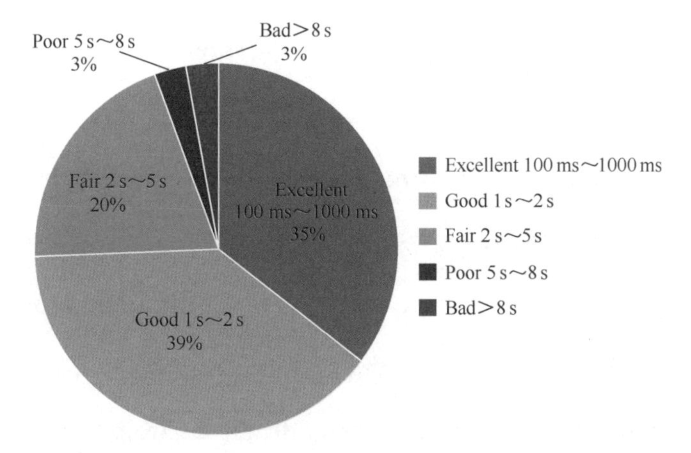

Source: China Video Service Users' Experience Standards Working Group

Figure 4-25 Distribution of initial video loading durations of Internet video services

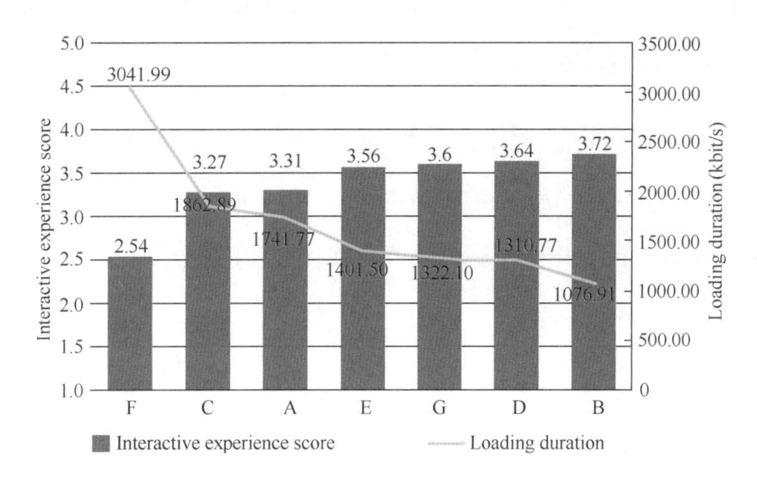

Source: China Video Service Users' Experience Standards Working Group

Figure 4-26 Interactive experience scores of different Internet video CPs

4.3.4 Evaluation of Viewing Experience of Internet Video Services

1. Frame freezing sometimes occurs during Internet video playback

12% of users encounter different levels of frame freezing during video playback. Once frame freezing occurs, it usually takes a relatively long time to resume playing. The average resumption duration is 6 s and the longest one reaches 20 s.

2. The Internet video viewing experience during idle hours is better than that during busy hours

According to the sample data analysis result, the Internet video viewing experience at 04:00 is the best and the average score is 4.47. From 05:00, the average viewing experience score starts to decrease. From 08:00 to 15:00, the average viewing experience score decreases significantly and the average frame freezing duration increases from 5700 ms to over 6000 ms. From 15:00 to 18:00, the viewing experience is relatively stable. At 20:00, the average frame freezing duration reaches 6400 ms and the viewing experience is the poorest of the day. After 23:00, the average viewing experience score starts to increase and reaches the highest from 03:00 to 04:00, and see Figure 4-27.

According to the analysis result of users' average video downloading speed, during busy hours, the number of users accessing video websites increases significantly as the number of online users increases. As a result, the networks and video service platforms congest and the valid bandwidth for each user to enjoy video services decreases. In this situation, the frame freezing occurrence probability increases and the video viewing experience score decreases, and Figure 4-28 indicates the details.

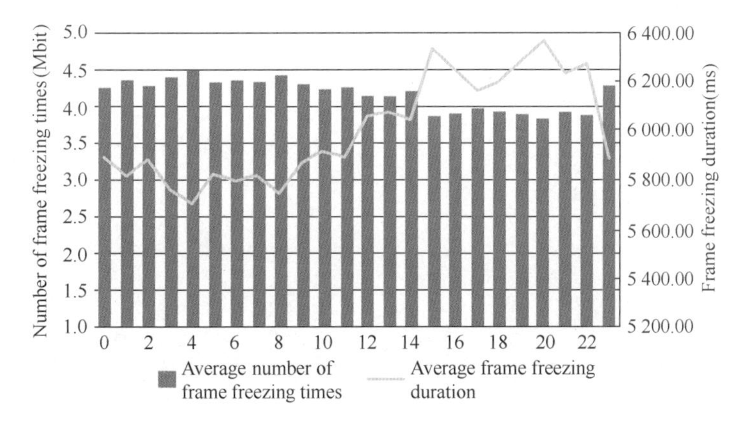

Source: China Video Service Users' Experience Standards Working Group

Figure 4-27 Internet video viewing experience in different periods of time

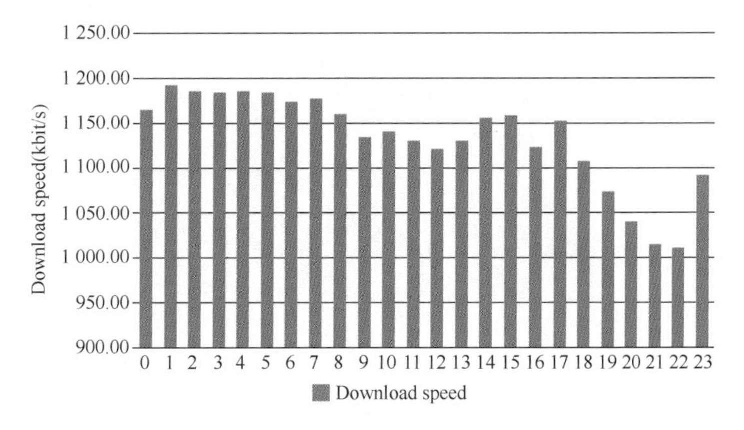

Source: China Video Service Users' Experience Standards Working Group

Figure 4-28 Internet video download speeds in different periods of time

According to the evaluation result, the frame freezing occurrence probability during Internet video playback mainly related to the available bandwidth. When user quantity increases but the total bandwidth does not increase accordingly, the viewing experience score decreases. Video CPs can increase the buffer to reduce frame freezing. For example, buffer sufficient data during initial video loading or after frame freezing ensures that users can continuously watch videos in certain time duration without frame freezing again. However, this can cause the decrease of interactivity scores.

5. Known Issues and Prospects

This activity, the first large-scale activity promoted by the industry, is conducted to evaluate real-time users' experience of various video services. It is of great value to set up a nationwide real-time users' experience evaluation system. This fills the gaps of related national and international standards. The evaluation shows the following outcomes.

- Because of the effective assurance provided by private network, the overall user experience of IPTV services is good and the viewing experience is excellent. However, the current network capability is insufficient for HD services, especially for 4K service. As the proportion of HD and 4K videos increases, E2E network optimization needs to be considered to improve the interactivity and viewing experience.
- Currently, users' experience of Internet video services has great variation.

With the bandwidth cost increases, video CPs proactively take measures to resolve the conflict between better video quality and more bandwidth expense. Video CPs also need to provide media sources with different video qualities to meet various users' requirements. As restricted by the network, the qualities of Internet videos are generally poor and the frame freezing occurrence probability during the video playback is relatively high. Therefore, various measures must be taken so that users can have continuously smooth video viewing experience on the network. Users' satisfaction of the experience will bring positive impact. Video CPs have done the best in interactive experience. Most Internet video CPs can ensure that the interactive operation is completed within 2 s and some can ensure that the operation can be accomplished within 1 s.

In China, Video Service Users' Experience Standards Working Group plan to carry out the following work later.

- Sum up the experience and extend evaluation of video service experience to all basic telecom carriers and other network carriers, broadcast and TV carriers, cable carriers, Internet video CPs, test scheme developers, and chip manufactures.
- Extend the evaluation to cover different networks including Wi-Fi, optical fiber broadband, cable TV, and 3G and 4G networks, different video services including IPTV, Internet TV, cable TV, and mobile video services, as well as more domains such as video surveillance, video communication,

and video conference.

- Extend the evaluation to cover third- and fourth-tier cities, rural areas, and distant areas so as to more thoroughly reflect the differences of video experience in different areas in China.
- Continuously extend the dimensions and systems for evaluation of video service experience. The dimensions include human-machine interactive experience and service usability, for example, remote control interactive experience, GUI usability experience, and voice identification experience.
- Start to implement network transformation and convergence so that the future network will be more suitable for video transmitted and can be applied to more rich service scenarios.

With the improvement of related techniques, users' expectations for video experience will gradually increase. The working group will continuously conduct survey and amend the video experience assessment standards so that users' real satisfaction with video experience can be achieved. The video experience assessment standards provide an important choice for industry development. It becomes a key means for carriers and Internet services to enter markets such as family service, social public service, and advanced manufacturing industry. It is also a weight for ensuring customer loyalty and lowing customer churn rate. In addition to cooperation with carriers and video CPs, cooperation with more third parties will be established. Measures such as user binding or cross-sales can be taken to gather audiences for the third parties.

The industry should start implementing network transformation to be more suitable for video transmission and more applicable to various service scenarios for the future network. The real-time service capability for service and network interconnection must be developed, which will finally contribute to the video industry development.

Acronyms and abbreviations

1080P	1920 × 1080 Progressive Scanning
720P	1280 × 720 Progressive Scanning
ABR	Available Bit Rate
AP	(Wireless) Access Point
APP	Application
CDN	Content Delivery Network
DLSP	Distance Learning Services Platform
DNS	Domain Name System
DRM	Digital Right Management
DSLAM	Digital Subscriber Line Access Multiplexer
DTS	Decode Time Stamp
DTV	Digital Television
EPG	Electronic Program Guide
FTTH	Fiber To The Home

HAS	HTTP Adaptive Streaming
HLS	HTTP Live Streaming (Apple)
HPD	HTTP Progressive Download
HTTP	HyperText Transfer Protocol
IGMP	Internet Group Management Protocol
IPTV	Internet Protocol Television
KPI	Key Performance Indicator
KQI	Key Quality Indicator
MDI	Media Delivery Index
uVES	User Video Experience Standard
OTT	Over The Top
PC	Personal Computer
PDA	Personal Digital Assistance
QoE	Quality of Experience
QoS	Quality of Service
STB	Set Top Box
STU	Set Top Unit
TCP	Transmission Control Protocol
TS	Transport Stream
VoD	Video on Demand
VR	Virtual Reality
HFC	Hybrid Fiber Coax